AF228467

WORMS

by Martha London

Cody Koala

An Imprint of Pop!
popbooksonline.com

abdobooks.com

Published by Pop!, a division of ABDO, PO Box 398166, Minneapolis, Minnesota 55439. Copyright © 2021 by POP, LLC. International copyrights reserved in all countries. No part of this book may be reproduced in any form without written permission from the publisher. Pop!™ is a trademark and logo of POP, LLC.

Printed in the United States of America, North Mankato, Minnesota

082020
012021

THIS BOOK CONTAINS RECYCLED MATERIALS

Cover Photo: Shutterstock Images
Interior Photos: Shutterstock Images, 1, 5 (bottom left), 10–11, 19 (bottom left), 19 (bottom right); iStockphoto, 5 (top), 5 (bottom right), 6, 13, 15, 16, 19 (top), 20; Steve Shinn/Alamy, 9

Editors: Christine Ha and Brienna Rossiter
Series Designer: Sophie Geister-Jones

Library of Congress Control Number: 2019954986
Publisher's Cataloging-in-Publication Data
Names: London, Martha, author.
Title: Worms / by Martha London
Description: Minneapolis, Minnesota : POP!, 2021 | Series: Underground animals | Includes online resources and index.
Identifiers: ISBN 9781532167652 (lib. bdg.) | ISBN 9781532168758 (ebook)
Subjects: LCSH: Worms--Juvenile literature. | Soil invertebrates--Juvenile literature. | Burrowing animals--Juvenile literature. | Underground areas--Juvenile literature.
Classification: DDC 592--dc23

Hello! My name is

Cody Koala

Pop open this book and you'll find QR codes like this one, loaded with information, so you can learn even more!

Scan this code* and others like it while you read, or visit the website below to make this book pop.

popbooksonline.com/worms

*Scanning QR codes requires a web-enabled smart device with a QR code reader app and a camera.

Table of Contents

Chapter 1

Wriggling Worm

An earthworm is a type of **invertebrate**. The body of an earthworm is divided into many **segments**. The worm uses them to move.

A worm can regrow some parts of its body if they get cut off.

Watch a video here!

Earthworms live on every
continent except Antarctica.
They tunnel through the dirt.
Most worms need to live
in wet, loose soil. This soil
makes digging easier. It also
keeps worms' bodies **moist**.

Chapter 2

Soft Body

An earthworm has a long, skinny body. Tiny hairs called **setae** cover its skin. The hairs help the worm grip the ground.

seta
Learn more here!

Earthworms have heads.

But they do not have eyes.

An adult worm has a band

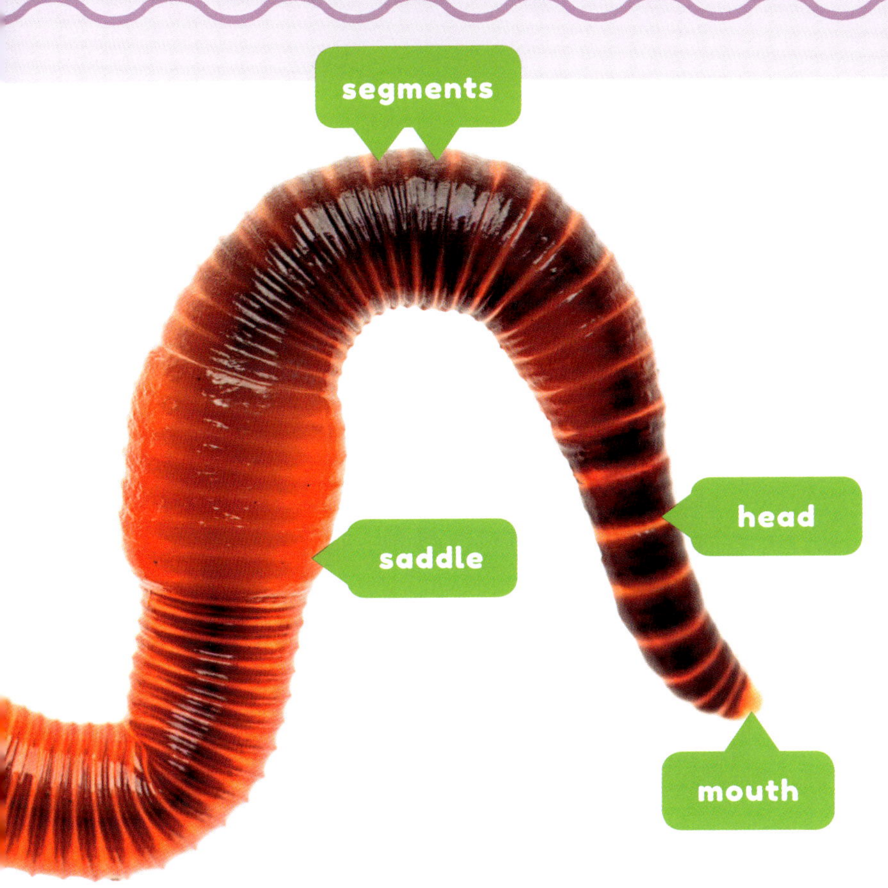

near its head. The band is called a saddle. The worm uses it to **reproduce**.

Each earthworm creates eggs. It makes a **cocoon** to protect them. Young worms hatch from the eggs. They look similar to adults. But young worms are much smaller.

Scientists found one earthworm that was more than 9 feet (2.7 m) long.

egg

Chapter 3

Under the Surface

Earthworms breathe through their skin. Their skin must stay wet. Sunlight can dry their skin out. So, worms avoid the sun. They spend most of their time underground.

Learn more here!

However, worms do come to the surface sometimes. They often come out at night. Or they crawl out after rain. But coming to the surface puts worms in danger. Birds may eat them.

Chapter 4

Good for the Ground

Earthworms are good for the soil. Their tunnels create holes that let rain get to plant roots faster. Worms also break up hard soil. Roots can grow more easily.

Complete an
activity here!

Worms eat as they crawl and dig. They eat dead leaves and dirt. Their waste puts **nutrients** in the soil. Then the soil takes in those nutrients. This process can make the soil richer. It can help plants grow.

Making Connections

Text-to-Self

Have you ever seen an earthworm? Where was it?

Text-to-Text

What books have you read about other invertebrates? How are those animals similar to earthworms? How are they different?

Text-to-World

Earthworms live on every continent but Antarctica. Why would it be hard for these worms to survive in Antarctica?

Glossary

cocoon – a protective covering.

invertebrate – a type of animal without a spine.

moist – slightly wet.

nutrient – part of food that humans, animals, and plants need to stay strong and healthy.

reproduce – to make new offspring.

segment – one small part of a whole.

seta – a tiny, stiff hair along a body.

Index

Online Resources

popbooksonline.com

Thanks for reading this Cody Koala book!

Scan this code* and others like it in this book, or visit the website below to make this book pop!

popbooksonline.com/worms

*Scanning QR codes requires a web-enabled smart device with a QR code reader app and a camera.